Secret Of Me Arithmetic: 70 .... Super Speed Calculation & Amazing Math Tricks

How to Do Math without a Calculator

By: Jason Scotts

TABLE OF CONTENTS

Publishers Notes... 3

Dedication 4

Chapter 1- Mental Math Arithmetic- Getting Rid of the Calculator
.... 5

Chapter 2- Secrets to Mental Arithmetic- 10 Ways to Add
Without a Calculator 9

Chapter 3- Secrets to Mental Arithmetic- 10 Ways to Subtract
Without a Calculator 13

Chapter 4- Secrets to Mental Arithmetic- 10 Ways to Divide
Without a Calculator 16

Chapter 5- Secrets to Mental Arithmetic- 10 Ways to Multiply
Without a Calculator 20

Chapter 6- Secrets to Mental Arithmetic- 10 Ways to Estimate
When Multiplying...24

Chapter 7- Secrets to Mental Arithmetic- 10 Ways to Estimate
When Dividing..28

Chapter 8- Secrets to Mental Arithmetic- 10 Ways to Check
Answers...31

About The Author..35

Jason Scotts

PUBLISHERS NOTES

Disclaimer

This publication is intended to provide helpful and informative material. It is not intended to diagnose, treat, cure, or prevent any health problem or condition, nor is intended to replace the advice of a physician. No action should be taken solely on the contents of this book. Always consult your physician or qualified health-care professional on any matters regarding your health and before adopting any suggestions in this book or drawing inferences from it.

The author and publisher specifically disclaim all responsibility for any liability, loss or risk, personal or otherwise, which is incurred as a consequence, directly or indirectly, from the use or application of any contents of this book.

Any and all product names referenced within this book are the trademarks of their respective owners. None of these owners have sponsored, authorized, endorsed, or approved this book.

Always read all information provided by the manufacturers' product labels before using their products. The author and publisher are not responsible for claims made by manufacturers.

© 2013

Manufactured in the United States of America

DEDICATION

This book is dedicated to my greatest source of support-my parents.

Jason Scotts

Chapter 1- Mental Math Arithmetic- Getting Rid of the Calculator

Mental calculations use only the human mind in order to find a solution to a mathematical problem. The human brain is used as a calculation tool, and the end results are determined from a mental computation of the arithmetic data itself. A calculator or computer is not used to find the end result of the arithmetic problem.

Techniques

Various mental techniques are frequently used to find a correct answer for a mental math calculation; these mental math techniques usually take advantage of the decimal and numeral systems. There are certain systems that may be used, and the method that is chosen is determined by the particular calculations that are being performed. A radix may be used for easy tasks that simply use a decimal point location to decide the correct answer for a math problem. Multiplying and dividing by 10 can be accurately done by moving the decimal point to the correct place within the numbers being used. A hexadecimal system needs to be used when the math problem uses other multiplying or dividing numbers. Using the number seven as a multiplier or divider may create a more complex mental math problem.

Techniques and Other Methods

There are other techniques and methods that may be used to find the correct answers to mental math arithmetic problems. The sum of the first nine digits of a group of nine numbers may be added, and this group of nine digits is called an operand. This technique uses the sum of the digits and several other operations to arrive at a correct answer. This particular method is called casting out nines. This type of mental math calculation may be used after a certain number of practice sessions in order to determine the correct sequence of operations needed.

Estimation and Mental Math Arithmetic

A second method that is used for mental math arithmetic is a technique called estimation. This type of mental math uses a necessary scaling of the numbers. This estimation will instruct the mental math user to estimate the number of digits and to be aware of the expected number of digits for the correct answer. Counting the number of digits for the final answer can assure the mental math processor that the answer is within an expected range and has the correct number of digits.

Factors Involved

The characteristics of the operands involved will help to determine various features of the final answer; multiplying two factors that each end in 5 will indicate that the correct answer should end in 5 as well. A number that is a multiple of two certain numbers may require that the answer be a multiplier of the two factors that were originally used. Prime numbers and prime factors help to determine the type of outcome for a set of multipliers as well.

Guessing an Answer

Guessing an answer may be used when certain amounts of information are known. A target number may be used in order to keep the final guess answer within a certain range of numbers.

Multiplying From Right to Left

This method of mental math arithmetic is often used when the digits are small. The first two digits on the right are multiplied together, and the remainder is carried over to the next set of two digits on the left. This technique is repeated until the final number is determined.

Multiplying By 5

This mental math arithmetic is a simple math calculation. The number is multiplied by 10 and then divided by 5. Mental math arithmetic uses this calculation frequently when the multiplier is a prime number such as 5.

Psychological Skill

Mental math arithmetic uses a certain amount of mental exertion and psychological skill in order to arrive at a correct answer. Mental and physical exertion must be exercised at a particular level in order to produce a certain performance level. Physical exercise tends to enhance mental performance as well. A mental work load may need a certain degree of physical ability in order for the performance to be correct and effective.

Perception and motor ability have been correlated. Over exertion physically has been determined to have a negative effect on cognitive skill level. There tends to be a certain balance of cognitive and physical exertion that needs to be reached in order to produce a maximum outcome of mental performance. This mental performance level is needed in order to correctly perform certain activities.

Summary

Mental math arithmetic is the calculation of numerical outcomes using only the brain as a calculation tool. A calculator or computer is not used. There are various ways to determine the correct outcome of a mental arithmetic problem by using several techniques and methods. These methods use multipliers and number sequencing. There are unique characteristics for some number factors and operands that may be used.

This special method can assist in finding the correct answer by using only the person's mind and psychological makeup. The psychological makeup of an individual may be influenced by the correlation between the person's mental ability and the person's overall fitness level.

CHAPTER 2- SECRETS TO MENTAL ARITHMETIC- 10 WAYS TO ADD WITHOUT A CALCULATOR

Among the fast learners, learning to add without a calculator is extremely easy. Although there are those people who are gifted in doing all forms of calculations without a calculator, there are those too who by regular practice can as well learn how to make this happen. This actually makes someone appear smart and be perceived as a genius. However, it may take time before one can be able to add off head and give an error free answer. The key to making this happen is by first mastering the small numbers like 0 to 5 before starting working with the bigger numbers like from 6 to 9.

One may also opt for the manual form of addition. This usually involves the use of a scribbling board or paper. One can then roughly draw pieces of sticks or any other symbols and manually calculate them. Whereby a lot of symbols have to be drawn, one can create bundles of each, either after every five or ten symbols so as to make addition easier. This is the very same way that young kids learn in school. The main disadvantage of this method is that it is time consuming.

Another option of adding without using a calculator is by taking advantage of the search engines. This is extremely easy for all that one needs to do is enter their search query, for example, 'what is fifty plus ninety', or rather, 'what is 50 + 90'. And, as soon as one presses the 'Enter' button, the answer is immediately displayed. To ensure that one gets the right number, it's best to look and compare the first answer with the others to ensure that they are all the same. Where there is doubt, one can try typing in the same query but on different search engines.

There also are the additional mathematical tables. These are easy to find in bookstores and though rarely used today, they can help one solve mathematical problems in an jiffy. Learning to use these tables is very easy and the great thing about them is that, they make the whole process of calculating even the very big and complex number quite easy. In addition to this, they contain a couple of tips on how to solve addition problems hence by making the process of calculation more effective and efficient.

Alternatively, one can choose to make use of the free online discussion forums and type in their addition problem. Since these discussion boards and forums are highly interactive, one gets to get their math problem solved very fast and easily once other members post in their answers or responses. Similarly, one can ask friends to help out through their social media channels like Facebook. It's obvious that everyone has a couple of smart friends, and at such a time, getting them to help out will be very efficient.

There also are various computer programs that can automatically do all the addition. A good example of such a program includes Microsoft office Excel which can indeed give the total number of things, figures or data in an instance. This is usually done by simply highlighting the numbers one wants to be added and just like that, MS Excel automatically displays the overall answer at the bottom. This alternative is most important when making calculations that involve the addition of big, complex and many numbers, especially when doing research or tabulating research findings.

For the many people who aren't fans of math, they can easily learn to add without having to use calculators by simply imagining the things they like. The logic behind this is that no one can be fooled with the things they like. If one loves money, then they would find it very easy to add numbers by imagining that its money they are adding. The same may apply in other instances too like imaging the number of pairs of shoes ones owns, or their special collections among others.

One may further embrace the use of patterned math blocks. The blocks are easy to work with especially for first time learners who are yet to master how to do addition math calculations. At such, they can manually count the blocks to enable them find the right answer to their math problems. This is only applicable if there is enough space to place the blocks and an environment that one is less likely to lose them.

The other option that one can use when adding without being required to use a calculator is through the use of their fingers. This is indeed one of the most commonly used formula for it is very easy and fast mainly because one can never go wrong using either their 10 finger or ten toes, or even the both of them.

Last but by no means least another alternative would be mastering numbers. The very same way that an elementary kid learns to master simple calculations such as 1 + 1 = 2 is the very same way that even an adult can. The beauty of calculations is that the numbers will always range from 0 to 9. At such, mastering these becomes very easy.

Chapter 3- Secrets to Mental Arithmetic- 10 Ways to Subtract Without a Calculator

Think of the first number in the equation in a + b format. Take 84 - 37 for example. First break down 84 into 79 + 5. The reason for this is because 79 is the closest number to 85 that can subtract 37 without regrouping. Now subtract 37 from 79 to get 42. You may be asking, "What happened to the extra 5 floating around?" Simply add the 5 to the difference and you have the difference to the original equation. So, there you have it, 84 - 37 = 47.

Think of the second number in the equation in the same a + b format. This technique is essentially the same as above but instead of adding the b, it is subtracted. Let's use the same equation, 84 - 37. Wouldn't the 37 be much nicer if it was broken down into 34 + 3? Well, there is nothing saying you can't do this! You now have 84 - 34 which is much easier to work with. Subtracting 34 from 84 you get the resulting difference of 50. Don't forget the 3 you separated from the 37. Now, just subtract the 3 from the 50 and we get the same difference as before, 47.

Round the second number down to an easier number to subtract from (preferably 0), then subtract again. Using the same example from above, 84 - 37, this would mean 37 becomes 30. First subtract 30 from 84 to get a difference of 54. The additional 7 would then be subtracted from 84 and again we get the same difference of 47.

Round the second up to an easier number to subtract from (again, 0 is normally easiest), then add. So, using 84 - 37, the 37 becomes 40. 84 - 40 results in a difference of 44. This is where the left over 3 comes in. Simply add the 3 and we get a sum of 47, as in the above equations.

Round the first number up or down to a simpler number; this principle is essentially the same as techniques 3 and 4. When you round the first number of the equation down, similar to the technique in the third option3, subtract instead of adding to the difference when you round up. The same is true for rounding down. When you round the first number of the equation up, like in technique 4, you round down you need to add to the difference instead of subtracting. Simply do the same thing to the difference as you did to the first number.

Visualize a number line in your head. Have the lowest number at one end, in this case 37 and the biggest number at the other end, in this case 84. 37 + 3 will get you to the nice round sum of 40. 40 to 80 can be seen as a block of 40 or, if you are more comfortable, you can group each set of 10 together to get 4 sets of 10. 80 to 84 then becomes an additional set of 4. Now let's look at the groupings we have divided the difference into 3, 40 (or 10 + 10 + 10 + 10) and 4. Add all groupings together to find your difference, 3 + 40 + 4 = 47.

Add the same number to both numbers in the equation to simplify the equation. Take 73 - 27 for example. If you add 3 to both numbers the problem becomes 76 - 30. Because you have added the same number to each number in the equation the difference of your new problem will be the same as the original. Clearly, 76 - 30 looks much simpler to work with than 73 - 27 (no regrouping). Now just subtract like you would any other problem. What is 76 - 30? It is 46, the same as 73 - 27.

Break up both numbers into smaller, simpler units. This method is best used for numbers that do not require regrouping. Take 99 – 43 then separate the tens and the ones places into two individual problems, 90 - 40 and 9 - 3. Simply solve both problems in your head (90 - 40 = 50 and 9 - 3 = 6) and add the differences of the two equations to get the difference of the original equation, 56.

Jason Scotts

Visualize the 100 square. Each row in the square is from 1 to 10 all the away to 100 (ex. 1 - 10, 11 - 20 and so on). Let's use 58 - 35. Separate 35 into 30 and 5. Start at 58 and go up 3 tens places to 28, then count back 5 units (not including 25). The number you land on is your answer, 23.

At this point you may be wondering how to easily subtract numbers with decimals. Again, simplify the equation; for example 45.58 - 13.45. Separate the whole numbers from the decimals to get two separate equations, 45 -13 and .58 - .45. Solve both equations individually using any of the examples above and add the differences.

CHAPTER 4- SECRETS TO MENTAL ARITHMETIC- 10 WAYS TO DIVIDE WITHOUT A CALCULATOR

Learning how to divide without a calculator can come in really handy, for instance, sometimes a test may not allow calculators, or when there is no calculator available, or simply because you do not need a calculator. There are various methods that can make dividing the numbers a little easier and avoid the need for a calculator. Learning the various ways of dividing without a calculator will be useful when dividing large numbers, decimals, or percentages.

One way of dividing can be by first estimating the numbers. When dividing a problem first take the larger number and round it closer to a number that is easily dividable. For instance, if the problem was 873 divided by 11, try estimating the 873 to 880 to help make the division process a little simpler and offer a way to check the final answer. The reason the estimation is to 880 is because 88 is easily divided by 11, so the answer is easily found.

A second method could be to estimate both numbers to the first digit. So, in the problem stated above, the 873 would be rounded to 900 and the 11 rounded to 10. This method can be tricky and does not always produce the most accurate of estimations, but it is an option when help is needed on a division problem.

A third method that comes in handy when dividing decimals is to take both the numbers and round them to the closest whole number. For instance, if the problem was 170.25 divided by 3.532, then you would notice the 2 and the 5, respectively. The first number would be rounded to 170, since the number to the right of the decimal point is less than 5; the number remains as its whole.

16

The second number would be rounded to 4, since the number to the right of the decimal is 5 or greater; it is rounded up one digit. So, now you are looking at 170 divided by 4. If you solve the problem without estimation, the answer is 48.2. If you solve the problem with estimation, the answer is 42.5.

A fourth method to divide without the use of calculators is when dividing decimals to first make both numbers whole. This is done by moving the decimals to the right of the number, but, keep in mind, that the decimal must be moved the same number of spaces in both numbers to keep the number equal to its original state. For instance, we will start with 3.532 since it has more spaces for the decimal to move in order to make it a whole number. The decimal has to move three spaces to make it 3,532. Now, you take 170.25 and move the decimal three spaces, adding a zero if there are no more numbers to fill in the space. So, in this case the number ends up being 170,250.

Now, you could take the two numbers in the previous problem and round them to the closest thousandth. The number 3,532 ends up being 4,000. The number 170,250 ends up being 170,000. Then, simply divide the numbers.

You could make it even easier and remove the same number of 0's from each number. The numbers would then be 170 divided by 4. Find the answer and add back the same number of 0's that were previously removed.

The seventh option for learning how to divide without a calculator is one that relates to percentage. Percentage is easier when we keep in mind that 100% is the entire amount, but breaking down that 100% can be easy as well if estimation is used. For instance, the problem is $18.72 is what percentage of $156.00? To solve this problem, you would take 18.72 and divide it by 156. You could round to the nearest tens making the 18.72 a 20, and the 156 can be rounded up to 160.

Another way is to complete the problem step by step. Try to break the problem into smaller steps. In the problem, 720 divided by 90 the numbers may seem large and daunting but take a second look. What if we drop the 0? Now the problem reads 72 divided by 9, which we know to be 8. When completing a problem in this manner, do not forget to bring back the 0. So the answer is 80. You could do this with larger numbers like 7,200 divided by 90 or even 900. Just keep in mind that what you do to one number you must do to the other to keep the problem unchanged.

There is always the good old fashioned paper and pencil method. Write the problem and divide it by finding how many times the divisor goes into the number to find the quotient. The quotient multiplied by the divisor must be the number being divided.

You could do the long division in your head by determining how many times the divisor would multiply into each number. This could get tricky when working with remainders and paper may be required to accurately solve the problem.

Jason Scotts

CHAPTER 5- SECRETS TO MENTAL ARITHMETIC- 10 WAYS TO MULTIPLY WITHOUT A CALCULATOR

In the past, people did not have calculators to be able to multiply. This meant that they had to find other methods for multiplying, or they simply would be unable to do multiplication. There are different methods that you can use in order to multiply without the use of a calculator. The great thing about mathematics is that there are various shortcuts that you can take to reach a quicker solution.

Outlined below are 10 ways to multiply without a calculator:

The Zero Trick: In order to do the zero trick, you have to have a larger number. The way that it works is you count up the zeroes and then add them up; For example, 25 x 5000 = 125000. The way that it works is you automatically put the zeroes at the bottom of the equation right away. This makes the numbers much easier to deal with, and simply, 25x5=125. Then add the three zeroes.

A Trick to Remember with Fives: The thing that you have to remember when you are multiplying fives is that they can only be two numbers. They will either equal a five or a zero. When five is being multiplied by even numbers, it will always equal zero, and when five is being multiplied by odd numbers, it will always equal a five. This is one method that you can use to reach a quicker conclusion when multiplying by a five. For example, 5x8=40 whereas, 5x7=35.

Write It down on Paper: One of the most common methods for multiplying without using a calculator is by writing it down on paper. The one disadvantage is that this is slower than doing it in your head, but it can help you to reach a definite conclusion quickly. This is a great method for the types of learners who need to see what they are doing in order to reach a conclusion. It makes it much easier to reach a solid solution without the fear of getting it wrong.

Practice Makes Perfect: If you want to get good at multiplying without the use of a calculator, then it is essential that you take the time to practice. It is not reasonable to expect to be good when you have not taken the time to practice.

Logarithm: Before calculators were used, a logarithm was commonly used. Logarithms were used up until the 1970s. At this point calculators became more affordable to the general public. Interestingly enough, logarithms were invented 400 years ago by the Scottish John Napier. They are the next best thing to reaching a solution without using a calculator.

Important to Know How to Add: If you want to be able to multiply larger numbers, then it is essential to have a full grasp of addition. Without a firm grasp of addition, it can be difficult to do multiplication of bigger numbers in your head. This is because it takes both multiplication knowledge and addition knowledge in order to do this with ease in your head.

Shortcuts Are Important: The great thing about mathematics is that there are numerous shortcuts that can be taken. You can even find and invent your own shortcuts. Mathematics is the type of field that has a broad opening, and there is always more that can be explored with it. It is important to look for shortcuts as well as constantly be trying to expand your knowledge of shortcuts.

Counting on Fingers: One of the things that you can do with addition when it comes to the bigger numbers with multiplication is to count the numbers up on your hands. This will lead to a faster solution.

The Nine Trick: How appropriate it is that the nine trick is method nine. The way the nine trick works is simple; when you are multiplying by nine, you take the number that you are multiplying on your finger and lower that finger. For example, if you were multiplying 3 x 9, you would lower the third finger on your left hand. This will show you a two and a seven, which equals 27. The only time that the nine trick will not work is when you are multiplying the number nine by a number that is large than nine. You can isolate the number in order to make it easier to multiply, but otherwise, you will be unable to use the nine trick.

Understanding the 11 Rule: When it comes to multiplying anything by 11 it is extremely simple to come to a solution; for example, 5 x 11 = 55. You are simply taking the five and multiplying it by one twice. You can do this with any number, so there is nothing that cannot be included.

These are 10 ways that you can multiply without the need for a calculator. The great thing about learning how to do mathematics without the use of a calculator is that it will increase your arithmetic skills and taking the time to learn these things will make you much better in the field that you are practicing.

CHAPTER 6- SECRETS TO MENTAL ARITHMETIC- 10 WAYS TO ESTIMATE WHEN MULTIPLYING

Multiplication can be difficult or simply take too much time when it needs to be done quickly. To increase speed and performance when doing mental math, learn to estimate effectively. In most real life situations, a quick estimation is better than a slow but exact answer. It isn't often that you need to know even one decimal place in everyday usage of math. Check out these 10 ways to estimate when multiplying and learn to use multiplication flawlessly in everyday life.

Relate to the Two, Three, Four, and Five.

Everyone has a few multiplication or "times tables" problems that stuck with them after grade school. A few commonly memorized ones include the twos, threes, fours, and fives. Most people are able to use the lower numbers with no problem. For example, two times three is six presents no problem. Six times two is twelve- also not a problem. So when a larger number shows up, such as 24, try to remember a simpler way to solve using smaller numbers. Six times what equals 24? Well 12 and 12 is 24 and six and six is 12 so by default four times six is 24.

Go Back to Basics, Then Up By Tens.

Using the same logic of sticking with small numbers, you can estimate a larger problem. For example, 305 times 198. How does this relate to a small number? Well 305 is close to 300, which is close to three mentally. 198 is close to 200, which is close to two mentally. Two times three is six; two times 300 is 600; 20 times 300 is 6000; 200 times 300 is 60,000. You have now estimated 305 times 198 effectively and much more quickly than with a pen and paper. The exact answer is 60,390.

Add A Bit and Subtract a Bit.

If you need a more exact answer to do a larger problem like 305 and 198, add or subtract a bit. In the mental math, 2 and 3 were used, however 305 is larger than 300 by five. 198 is only smaller than 200 by two. One can guess that the estimate will be smaller than the true answer. If you know that this is going to happen, account for it by adding a little. If the numbers used in mental math were both larger than the original exact numbers, subtract a little. This is not usually necessary, but results in a closer estimation.

Think of a Problem You Know Well.

Many people remember that six times six is 36, simply because it rhymes. Many also remember that 12 times 12 is 144, simply because it is the final part learned in grade school. If you know 12 times 12, you can find 13 or 11 times 12 by subtraction. This is much faster than writing it out by hand. If you know a single multiplication by heart, you also know the one above and below it, via addition and subtraction. Six times six is 36, so seven times six is 42 and five times six is 30.

Go for Tens and Fives

How can a person figure out five times 800? Simply find 10 times 800 and then divide by two. 8000 divided by two is 4000. Thus the answer becomes simple. If something is 9, 11, 4, or 6 and the answer does not need to be exact, just use five and 10. If you want a more correct answer, utilize the addition and subtraction method outlined above. This can be used in a variety of situations, including when decimals enter the equation.

Ignore decimals

Many people get caught up trying to figure out a decimal problem. In real life, it is usually obvious after the problem is finished, where the decimal should be. If you have a purchase of $5.50 with a six percent tax (.06), just ignore the decimal until the problem is finished. It will be obvious where the decimal should be placed in the end. Dealing with money also ties into a second estimation method.

Add Smaller Problems Together to Finish a Larger Problem.

Continuing with the example of $5.50 and six percent tax; 550 times six means a mental number of five. Five times six is 30, 50 times six is 300; 500 times six is 3000. The number $5.50 is composed of 500 and 50, so you can add those two problems together to get the answer. 500 times six is 3000, 50 times six is 300. The answer without a decimal is 3300. Now, where does the decimal go? Which sounds more like six percent of $5.50: $330.0 or $33.00 or $3.300 or $0.3300? Correct- 33 cents.

Get an Exact Answer for the Largest Problem, and Then Guess the Smaller Ones.

In the previous example, we had $5.50, but what if it were an annoying number like $5.37? Well, the most important part is the five dollars. Since you can easily find that six percent of $5 is 30 cents with the method above, you've already finished most of the problem. If an exact answer isn't needed- just guess the smaller added amount or use the five and ten method and simply assign it an easier value which will end up being close to correct.

Use Division

If you can do a bit of division, multiplication becomes much easier. The most difficult number can be broken down into a series of smaller numbers. Doing this with tens and fives was partially explained above, but it can be done with any multiplication problem. If nines are a weakness, divide the whole problem by three. At the end, multiply the answer by three. It simplifies the process if there is a certain multiplier which seems more difficult.

Visualize the Problem

This helps with division. Visualizing the things you want to divide or multiply helps many people to see the not-so-obvious solutions or ways to make the problem easier. Multiplying by seven is annoying to many, but when visualized it is easy to see that seven is almost like five, plus half again. So to estimate it, it will be possible to use five and then add another half to the answer or use the addition and subtraction method from above.

Multiplication is a basic skill that many people forget partially or entirely by the time they really need it. Knowing how to quickly estimate a product in your head is a great skill that will last a lifetime.

CHAPTER 7- SECRETS TO MENTAL ARITHMETIC- 10 WAYS TO ESTIMATE WHEN DIVIDING

Dividing is a skill that takes some practice. In certain cases, you might have to estimate as opposed to figuring out the answer. What are 10 ways to estimate when you are dividing figures?

Round the Numbers

In most cases, dividing round numbers is a lot easier than dividing other figures. For example, maybe you need to divide 224 by 17. Instead, you could divide 220 by 20. For figures that are five and over, you should round up. When the number is less than five, you want to round down. Doing so will give you the most accurate estimates. Practice rounding numbers. That way, when you have to divide, you can do it quickly.

Eliminate the Change

Maybe you are often put in charge of dividing up the bill when you go out to dinner with a group of friends. This is fine and dandy, but you are quite tired of having to crank numbers on the calculator! When you are figuring out the bill, eliminate the change; just round up. If you are dividing $322.59 by 10 people, it's a lot easier to just say that you are dividing $323 by 10 people. Chances are no one will mind putting in the extra cents.

Choose an Even Number

We often have a lot more ease diving even numbers than we do odd numbers, so you could also "round" the numbers in that way. Maybe you are trying to decide 357 by two. Opting to divide 358 by two will make the process a lot more efficient. Then, you will have a basic idea of what the answer is. When you change to even numbers, you usually wind up with an answer that is quite close to the correct one.

Choose Two Even Numbers

You need not choose only one number to make even when you are trying to estimate as you are dividing. If the numbers in the previous example were 357 and three, you could round to 358 and two. No, the answer is not going to be as accurate as if you just rounded one of them. However, it will still be pretty close, and after all, you are just estimating.

Think Backwards

Let's say that you have to divide 12 by four. If you are better at multiplication than division, you might want to say, "Well, what number do I need to multiply by four to get 12?" This strategy is an excellent one for people who think in that manner, and you can really sharpen your division skills by employing it on a regular basis.

Drawing a Quick Picture

Sometimes, visually seeing the numbers before you helps you to determine what the answer to the math problem is going to be. Sketch a quick picture either on a piece of paper or in your mind, and you might be able to see the actual answer. Remember, sometimes estimating will provide you with exactly the right answer because your methods are so incredibly accurate.

Visualizing the Results

In certain cases, people do not even need to draw pictures. They are just able to visualize the numbers and make them work in your head. For example, perhaps every time you have to divide eight by a number, you think of a pizza pie. You know that if you have one fourth of the pie, you have two slices. As a result, eight divided by four must be two. This technique works especially well when visualizing items that come in packages or slices.

Use Whole Numbers

As of now, we have been discussing whole numbers only. However, in some cases, you might have to divide fractions or decimals. Dividing these figures can definitely be challenging, but it is certainly not impossible; round up to or down to the nearest whole number for both of the digits. While this will not give you an extremely accurate sum, you will still be in the ballpark.

Long Division

Many of these tactics work well for short division, but what can you do when a long division problem arises? Perhaps you have to divide 253 by 42. Again, you can use rounded numbers for both of these. If you decide to divide 250 by 40, you are going to have a lot less difficulty than with the first answer. If you are in a course, the teacher might ask that you show work.

Ask a Friend

Maybe you are having a really difficult time with your numbers and dividing is just not your thing at all. When you are really in a bind, ask a friend for a clue. Your friend might say that the answer is between 10 and 20. At the very least, this range will allow you to focus. You no longer feel as though the answer could be any number in the entire world. You know that it is some reasonable sort of figure that falls into a particular span.

Learning how to divide is a skill in and of itself. When you need to estimate the answer, these tips are going to really help you succeed in the endeavor.

Chapter 8- Secrets to Mental Arithmetic- 10 Ways to Check Answers

The best way to check answers is definitely through the use of books. This is because all of the questions under the globe have answers most of which can be found in published works of literature. Books are beneficial as they usually carry information that can be traced to thousands of years ago. At such, one can check for answers of things that existed centuries ago by simply searching on books. Due to the modern day technology advancement, books have become much easier to use in checking answers mainly because one can do so much faster through the use of soft copy books or rather the e-books.

Search engines have a wide collection of information making it very easy for one to check answers through these avenues. One of the main reasons why checking answers via the search engines is reliable is because search engines carry a wide collection of information from all around the universe. As a result, checking answers from even the remotest and farthest parts of the world is made extremely easy. This therefore saves on a lot of things, like the cost of time, energy and even money.

The availability of online libraries is also an alternative for checking answers. Unlike the mainstream libraries, the online libraries are more user-friendly as they save on time. Furthermore, it is much easier to get answers to other related or similar answers for whatever one is searching for. Most of the online libraries can further be accessed for free and can retrieve answers from even the rarest sources, such as institutional data bases or other individual or institutional published works that cannot be found anywhere else.

Similarly, one can check answers through the use of experts in a particular field or even academicians. These two groups of people are widely knowledgeable on almost everything that is under their field of specialization. Experts and academicians can therefore be used in instances where one is looking for answers in a particular field meaning that, they are looking for more specific rather than the general answers. However, finding them could be quite a hassle for checking answers from them will usually demand a one on one form of communication.

When one is checking answers that cover a particular broad area, a focus group seems to work perfectly. This is because through the focus group, one gets an opportunity to get different answers, which in many cases are all right, but from different sources hence capturing different perspectives. Furthermore, focus groups make it possible to check answers that are out of the ordinary and ones that are influenced by different worldviews and people's perceptions. Such groups are very good especially when checking answers about something that is still new and interesting.

Another alternative to check answers would be by working with the mistakes. By starting to work with mistakes, it becomes easier for one to eliminate the assumptions hence get to check for the real and more accurate answers much easily. The more mistakes are eliminated, the easier it becomes to check answers.

Hypothesizing also could help in checking answers. The beauty of a hypothesis is that one is usually presented with an opportunity to work with various assumed expected answers. As a result, it becomes even much easier for one to work with a couple of hypothesis and out of these, they can check and come up with the most perfect answer or, the top best answers.

Almost all of the answers that need checking have already been worked on by someone else. Sampling and going through various research papers which are related to the topic will easily enable one check whether the answers are right or wrong. Since research papers are critically done and analyzed, it is more than explicit that the answers contained in there are the best working answers that there are available at the time.

Since not everyone can read, or rely too much on written works to check for their answers, and actually, many people are lazy at reading, then, listening to recordings is an alternative. Recordings are much easier to listen to and further saves on the time of checking for one can easily listen to specific parts of the tape to get to the exact information that they are looking for. Other than the recordings being audio, there are some which could be audio visual thereby making it easier for one to check answers even in visual forms.

Doing experiments could further help in ascertaining the validity of answers. This is mostly in areas that involve science related disciplines whose answers are not as easy to check. Due to this, one is expected to carry out an experiment that is related to the topic at hand. The findings got after the experiment are then used in checking answers. In the real sense, this is one of the most reliable ways to check answers.

ABOUT THE AUTHOR

Jason Scotts is an individual that loves to find solutions to problems. This is something that he has been encouraged to do all his life and his parents have supported every academic challenge that he has opted to take on. He is aware that solving mathematical equations may be simple to some and extremely difficult for others and aims to provide simple solutions for those who have challenges.

He had his own challenges as he would always grab the calculator to do even the simplest mathematical task. After a while he made the decision that he had to get out of this method of functioning as the day would come when he would not have access to a calculator and then what would he do.

As such he found some quick and easy techniques to solve mathematical problems. It is nothing confusing and really makes things that much easier to do. It also saves on the time spent looking for that calculator in the first place.

Printed in Great Britain
by Amazon

26688168R00030